Child-Centered Math

Logical Reasoning and Probability

35 Hands-On Activities

Grades 2–3

Written by Cindy Barden
Edited by Janet Bruno
Illustrated by Terri Sopp Rae
Project Director: Carolea Williams

Table of Contents

To the Teacher

The child-centered activities in this book will introduce your students to the concepts of logical reasoning and probability through a variety of hands-on projects. Activities are designed to be fun and educational, to reinforce and expand existing knowledge, and to provide students with ways to use logical reasoning and probability in a math context, across the curriculum, and in everyday life.

Logical Reasoning and Probability is designed as a handy resource for teachers, not a prescribed continuum of activities. Integrate the activities into your current mathematics program, keeping in mind the special needs of your students.

Have fun watching your students get excited about math as they participate in these interactive activities.

Getting Started

Most activities can be implemented in small- or large-group settings, but some are best suited to learning centers where a few students can work independently. When planning the amount of adult guidance or participation needed, keep in mind the materials to be used, the type of work space needed, and the activity level of the project.

Logical Reasoning and Probability is ideal for involving teacher aides or parent volunteers in the classroom. The directions are simple and easy to follow, and students will quickly become engaged in the activities. It may be helpful to place a laminated copy of each activity in a box with the materials. This allows for instant setup and cleanup.

The activities call for a combination of inexpensive, everyday objects and commonly-available commercial math manipulatives. As you review the materials lists, feel free to substitute materials as needed. A parent letter is included on page 36 to help you obtain various consumable materials.

About Logical Reasoning

Logical reasoning is an important life skill that helps make sense of the world around us. As students participate in logical reasoning activities, they learn to sort information and think through problems step by step. Encourage them to talk and write about their thought processes so they become conscious of effective problem-solving strategies.

About Probability

Probability is the chance or likelihood that something will happen. In math, probability can be expressed as a ratio or percentage. Students become familiar with probability at a young age. They play games of chance using dice and spinners, and they use weather forecast information to dress appropriately.

The activities in this book help students better understand how probability works. By playing games and recording results, they learn to make generalizations and then refine their predictions the next time. They also develop a sense of fairness as they examine the probability of winning games.

Compare and Contrast

Activity 1

Materials

- assorted objects
- paper
- pencils

Procedure

1. Comparing and contrasting items are good ways for students to use reasoning skills. Divide the class into partners. Ask each student pair to select two objects.
2. Ask students to list at least three ways the objects are alike and three ways they are different.
3. Have students share their lists in small groups. See if other group members can add to the lists.

Extension Compare and contrast two objects using a Venn diagram.

Notes:

Extensions:

First, Second, Third

Materials

- paper
- pencils

Procedure

1. Ask students to select a simple task such as brushing teeth, sharpening a pencil, or drawing a triangle.
2. Have them write, in order, all steps needed to complete the task. Remind them not to skip any steps, even if they are obvious. For example, before they can sharpen a pencil, they must find a pencil that needs sharpening.
3. Ask students to read their steps to the class.
4. Have the rest of the class listen carefully and point out any missing steps.

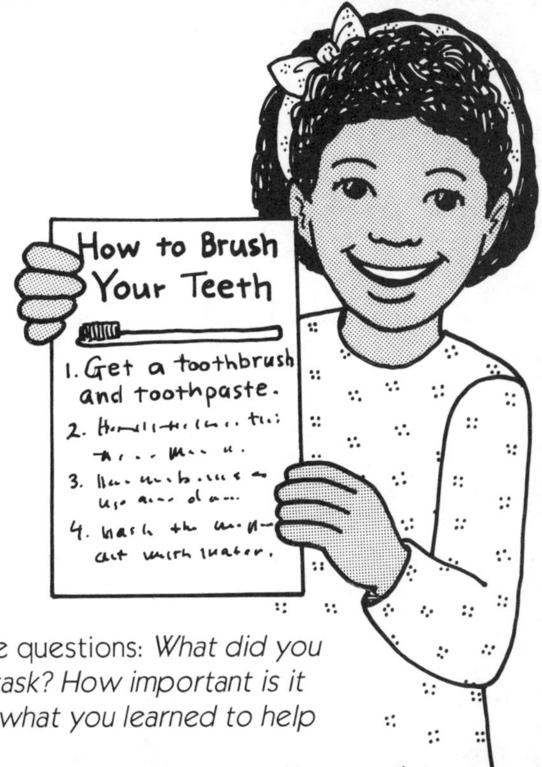

Extension Ask students to discuss these questions: *What did you learn about writing steps to complete a task? How important is it to include every step? How can you use what you learned to help solve math problems?*

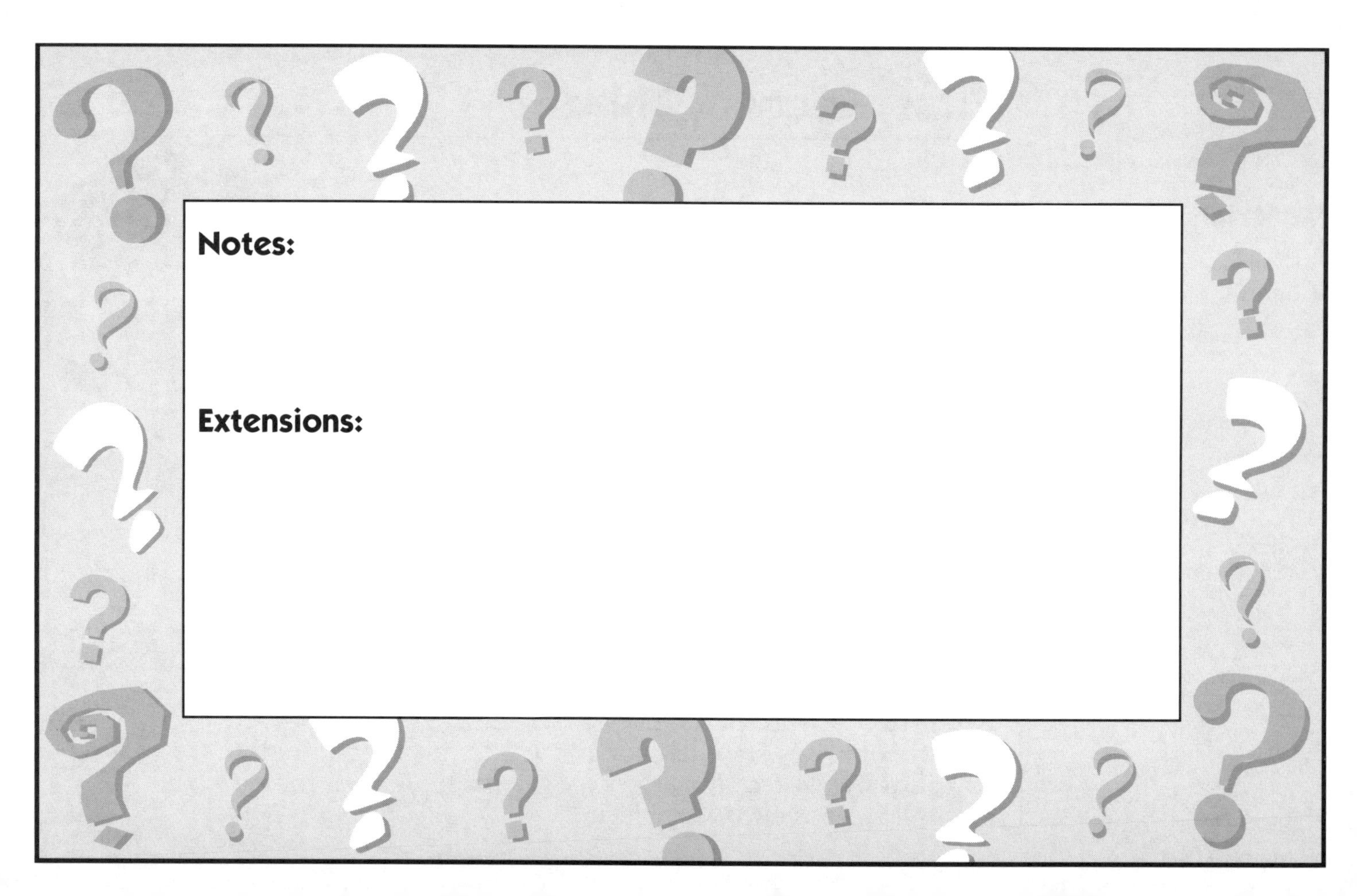

Notes:

Extensions:

Toothpick Triangles

Activity 3

Materials

- flat toothpicks
- paper
- pencils

Procedure

1. Give each student 12 toothpicks. Ask students to arrange the toothpicks to make as many equilateral triangles as possible without breaking or bending any toothpicks. Tell them they must use all 12 toothpicks.
2. When finished, have students draw the toothpick arrangements and write the number of equilateral triangles formed.
3. Have students compare their patterns with classmates' patterns to see which arrangement made the most triangles.

Extension Have students use 12 toothpicks to form five squares. Hint: The squares can be different sizes.

Notes:

Extensions:

"X" Marks the Spot

Materials

- paper
- pencils

Procedure

1. Ask each student to write clear directions from your classroom to some place in the school, such as the office or library. Tell students not to name the place—they can call it "Point X."
2. Have students trade directions with a classmate and attempt to follow the directions exactly as written. If the directions are not accurate, ask students to rewrite them.
3. Have students summarize what they learned about giving clear directions.

Variation Following their partner's directions, have each student draw a map from the classroom to Point X.

Notes:

Extensions:

Number Cards

Activity 5

Materials

- 3" x 5" index cards
- 12 chips (game pieces, buttons, pennies)
- small container

Procedure

1. Working with a small group, have students write the numbers *0–9* on index cards, one number per card. Place the chips in the center of the table.
2. Have one student gather the cards, shuffle them, and deal three cards to each student.
3. Ask students to arrange their three cards to make the largest possible number and lay it face up on the table. The student with the largest number takes one chip.
4. Have students choose three new cards. Shuffle and reuse cards if necessary. Have students play until all chips are gone. The student with the most chips at the end wins.

Variation Have students make the smallest possible number using three cards.

Extension Deal four or more cards per student and play as described above.

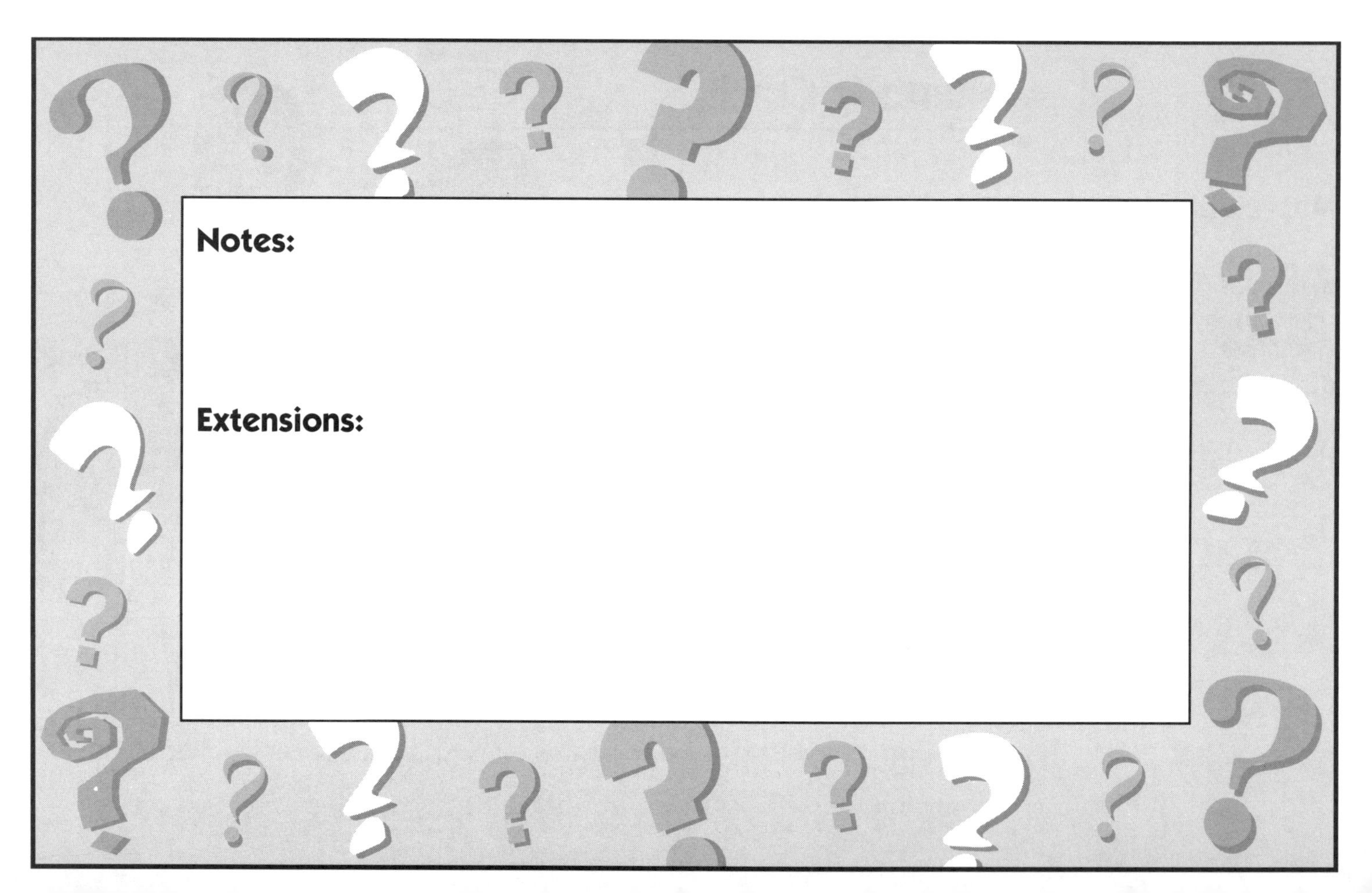

Notes:

Extensions:

Who Am I?

Activity 6

Materials

- paper
- pencils

Procedure

1. Have students write six to ten clues about themselves, going from very general to specific.
2. Model this procedure with clues about you. For example, "I'm a woman. I live in California. I'm a teacher and a mom. I teach at Arroyo School. I have brown hair. I'm 5'4" tall."
3. Collect student papers and read one or two papers each day, pausing after each clue.
4. See how many clues it takes the class to guess the right person.

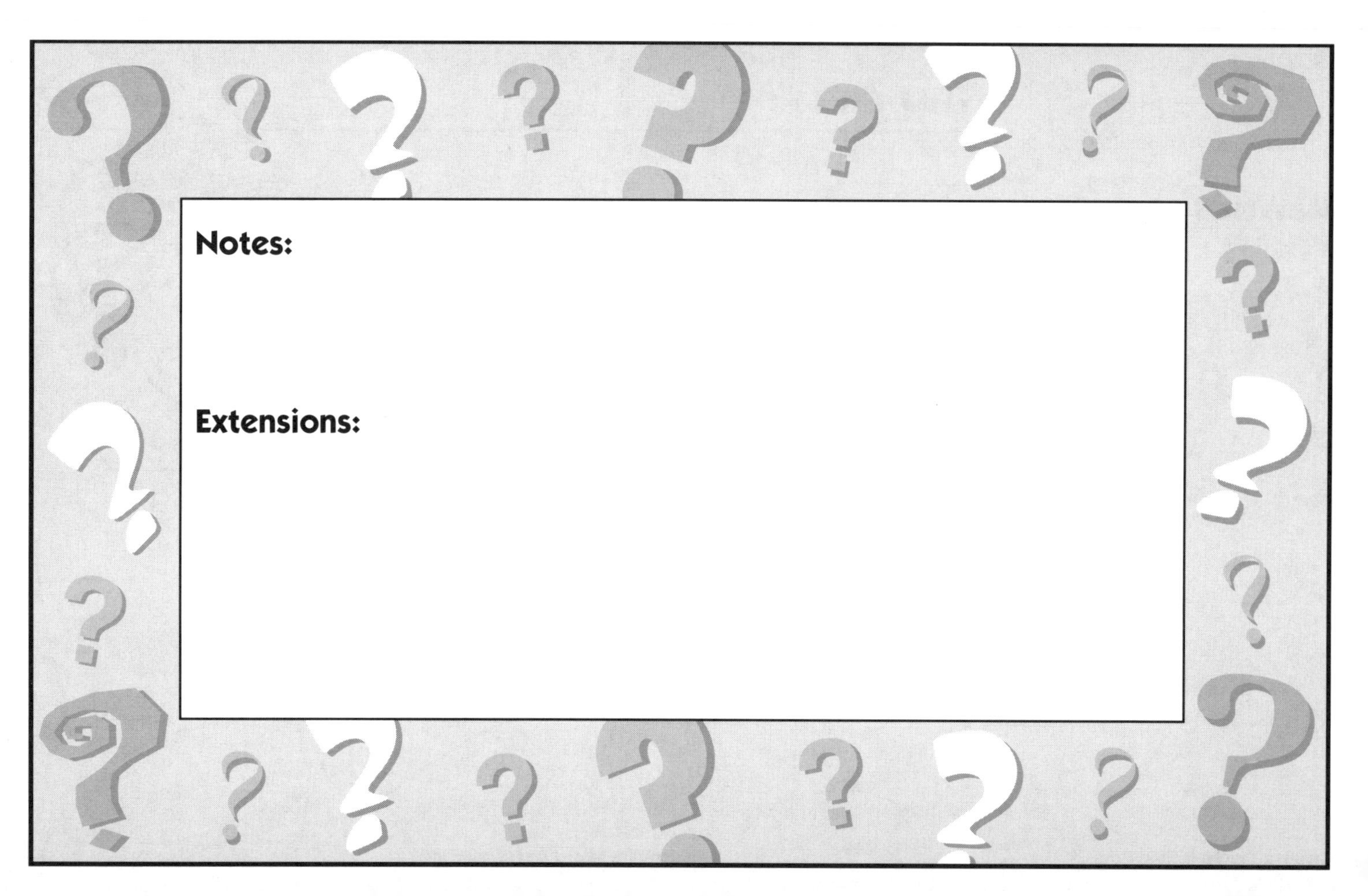
Notes:
Extensions:

Tangled Triangles

Materials

- drawing paper
- pencils
- rulers

Procedure

1. Have students make six dots on their paper with three dots in each row, as shown. Each dot should be about two inches apart.
2. Ask students to draw straight lines to connect the dots in any way they wish, making as many triangles as possible. Lines may cross each other more than once.
3. Have students count the number of triangles they made. Remind them to count small triangles and larger ones formed by two or more small ones.

Extension Repeat the activity using nine dots.

Notes:

Extensions:

Comic Codes

Activity 8

Materials

- newspaper comic strips
- scissors
- pen

Procedure

1. Cut out comic strips from the newspaper. Replace the dialogue with words spelled in code (e.g., A = •, B = ?, C = @).
2. Give each small group of students one comic strip. Have them use picture, syntax, and punctuation clues to break the code.

Notes:

Extensions:

Take a Turn

Activity 9

Materials

- small paper plates
- beans
- paper
- pencils

Procedure

1. Give each student pair a paper plate and 21 beans.
2. Have students place all 21 beans on the plate.
3. Tell each player to take turns removing either one or two beans from the plate.
4. Whoever forces the other player to take the last bean wins the game.
5. Ask students to keep track of the number of games they win.

Variation Start with 51 beans, and allow players to remove from one to six beans at a time.

Notes:
Extensions:

Guess My Number

Activity 10

Materials

- chalkboard
- chalk

Procedure

1. Pick a two-digit number (digits should be different).
2. Have students guess the number. Each time they guess, record the guess with the following code:
   - ● = no correct digits
   - ✓ = correct number, but not in the correct spot
   - ★ = correct number in the correct spot
3. Encourage students to explain the reason for each new guess. For example, "When Randy said 27, we had two correct digits, but they weren't in the right spot, so I reversed them to get 72."

Guess
23 ✓
46 •
82 ★
87 ✓
27 ✓✓
72 ★★

Extension Extend the number to three digits.

Notes:
Extensions:

Line Up the Books

Activity 11

Materials

- 3 books
- sticky notes
- markers
- chalkboard
- chalk
- 3" x 5" index cards
- drawing paper
- pencils

Procedure

1. Number the three books *1, 2,* and *3* with sticky notes, and line them up in a row.
2. Ask volunteers to demonstrate other ways the three books can be arranged. Show students how to record each arrangement on the chalkboard. Continue until students have found all six ways.
3. Give each student pair four index cards. Ask students to number the cards *1, 2, 3,* and *4.*
4. Tell students the cards represent four books. Ask them to manipulate the cards to find all possible ways the books can be lined up on a shelf.
5. Have students record each arrangement, working systematically.

Notes:

Extensions:

Mix and Match

Activity 12

Materials

- 3 winter hats
- 2 scarves
- drawing paper
- colored pencils or markers

Procedure

1. Place materials at a center. Ask students to figure out how many different combinations they can make with three hats and two scarves.
2. Encourage students to assemble all possible combinations.
3. As they work, ask students to draw pictures or diagrams to represent the combinations they made.

Extension Have students complete this activity using three articles of clothing that can be worn together, such as two hats, three sweatshirts, and three pairs of jeans.

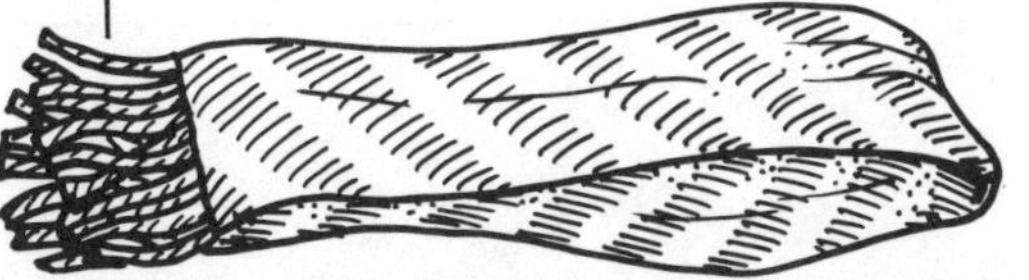

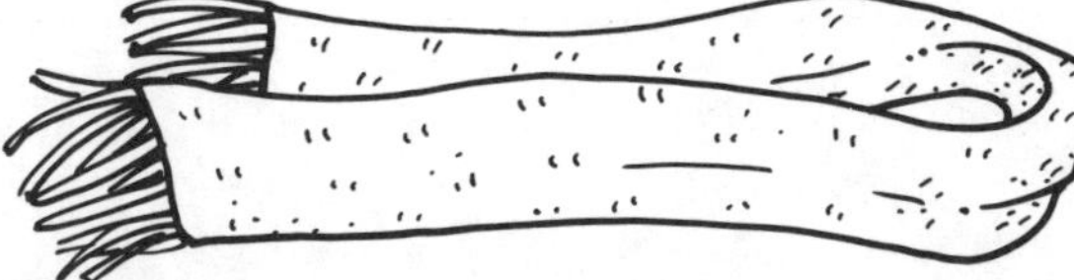

Notes:

Extensions:

Parking Problem

Activity 13

Materials

- red, white, and blue counters (or pieces of construction paper)

Procedure

1. Divide the class into partners and give each student pair an assortment of red, white, and blue counters.
2. Ask students to use counters to solve this problem: *Ten cars are parked in a row. The middle two are blue. The car on the far right is blue. The last two cars on the left are red. The rest of the cars are white. How many white cars are there and where are they parked?*
3. Ask students to demonstrate how they arrived at their answers and compare different strategies.
4. Make up similar parking problems for students, or have small groups of students write their own. Show students how to physically arrange the "cars," and then write the clues.

Notes:

Extensions:

Kittens, Kittens, Kittens

Activity 14

Materials

- drawing paper
- markers
- scissors
- paper
- pencils, crayons

Procedure

1. Tell students to make their own paper manipulatives to help solve this problem: *There are five kittens for sale—one black, one white, one striped, one gray, and one orange. How many possible combinations of kittens are there if someone wants to adopt two?*
2. Ask students to use their manipulatives to solve the problem.
3. Have them draw diagrams or pictures showing the different combinations of two kittens someone could select.
4. Have students compare their answers with a classmate to see if they have found all possible combinations.
5. Ask students to explain the methods they used to find the combinations.

Notes:

Extensions:

Making Sundaes

Materials

- counters
- drawing paper
- crayons
- pencils

Procedure

1. Tell students to solve this problem: *How many different ways can you make a sundae with two different ice-cream flavors and two different toppings?*
2. Let students pick the two flavors and the two toppings (e.g., nuts, whipped cream, fudge, butterscotch), and have them record their answers.
3. Ask students to share their strategies. Did they use counters or pieces of colored paper as manipulatives? Did they draw pictures of the combinations? Did they record with words?

Extension Increase the number of ice-cream flavors and/or toppings.

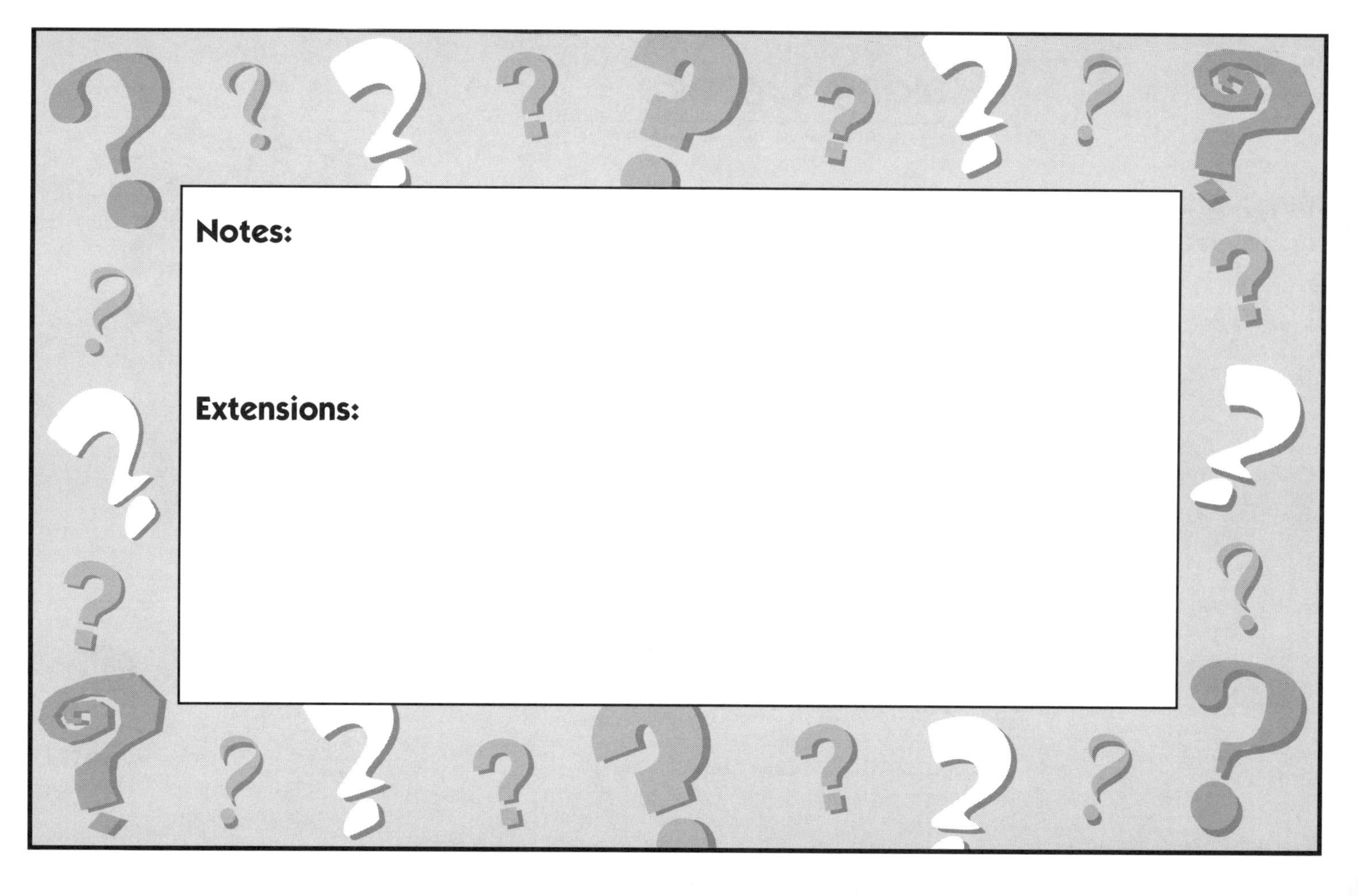
Notes:
Extensions:

A Dollar's Worth

Activity 16

Materials

- coins (pennies, nickels, and dimes)

Procedure

1. Working with three or four children at a time, ask students to make as many groups of coins worth 25¢ as they can. Each group of coins must be different (e.g., two dimes and one nickel, 25 pennies, five nickels).
2. Continue the activity, having students make as many coin combinations as possible to total 30¢, 50¢, 75¢, and $1.00.

Extension Have students devise a way to record different coin groups.

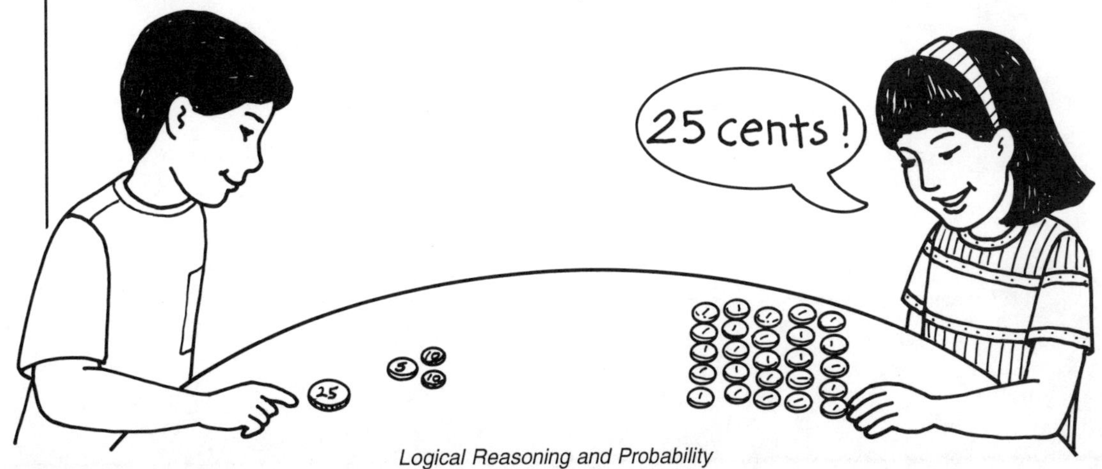

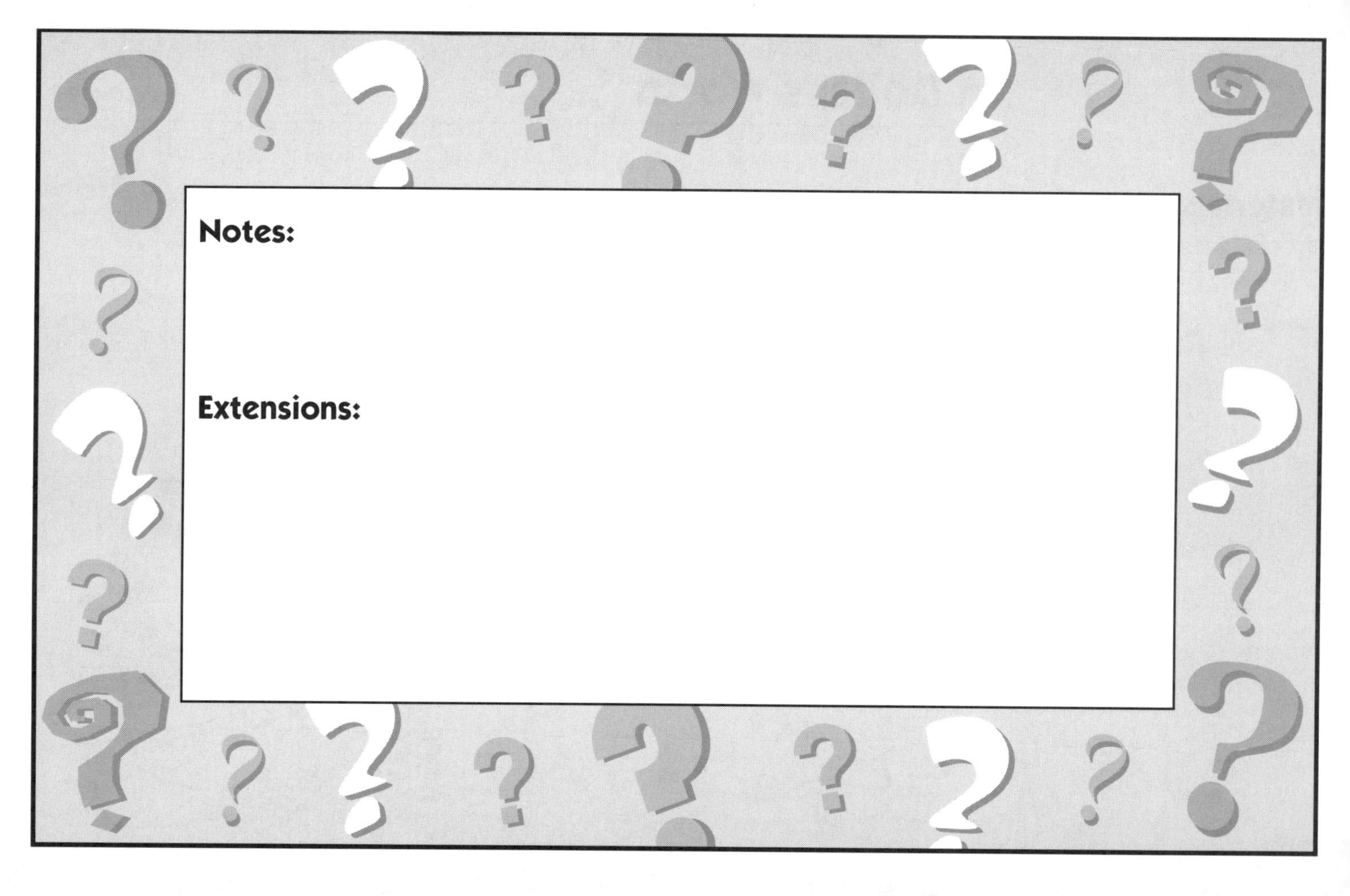

Notes:

Extensions:

Fill and Win

Activity 17

Materials

- 1" graph paper
- scissors or paper cutter
- counters or beans (two colors)
- paper
- pencils

Procedure

1. Cut the graph paper into 3" x 6" sections, and give each student pair one section and 30 counters, 15 of each color.
2. Have each player, in turn, place one or two counters on the grid. If two counters are placed, they must be in adjacent squares. The player who places a counter in the last square wins. The winner goes first in the next round.
3. Have students play several rounds and keep track of their scores.
4. Ask students if this is a fair game. Do both players have an equal chance to win? As a class, discuss winning strategies.

Notes:

Extensions:

Mystery Word

Activity 18

Materials

- chalkboard
- chalk

Procedure

1. Choose a three-letter word (e.g., *map*), but don't reveal it to the class. Draw a star on the board to represent the mystery word.
2. Have students take turns guessing the word. If the student's word comes before the mystery word in alphabetical order (e.g., *dog*), write the word above the star. If it comes after the mystery word (e.g., *tag*), write it below the star.
3. Discuss with students how these clues help narrow the possibilities.
4. If students guess the wrong word but have the correct initial letter (e.g., *mat, mop, met*), use the same strategy to help them discover the mystery word, alphabetizing by the second letter.
5. Continue to record students' guesses in alphabetical order until someone guesses the mystery word.

Extension Have students play the game in small groups or partners with one student taking the teacher's role.

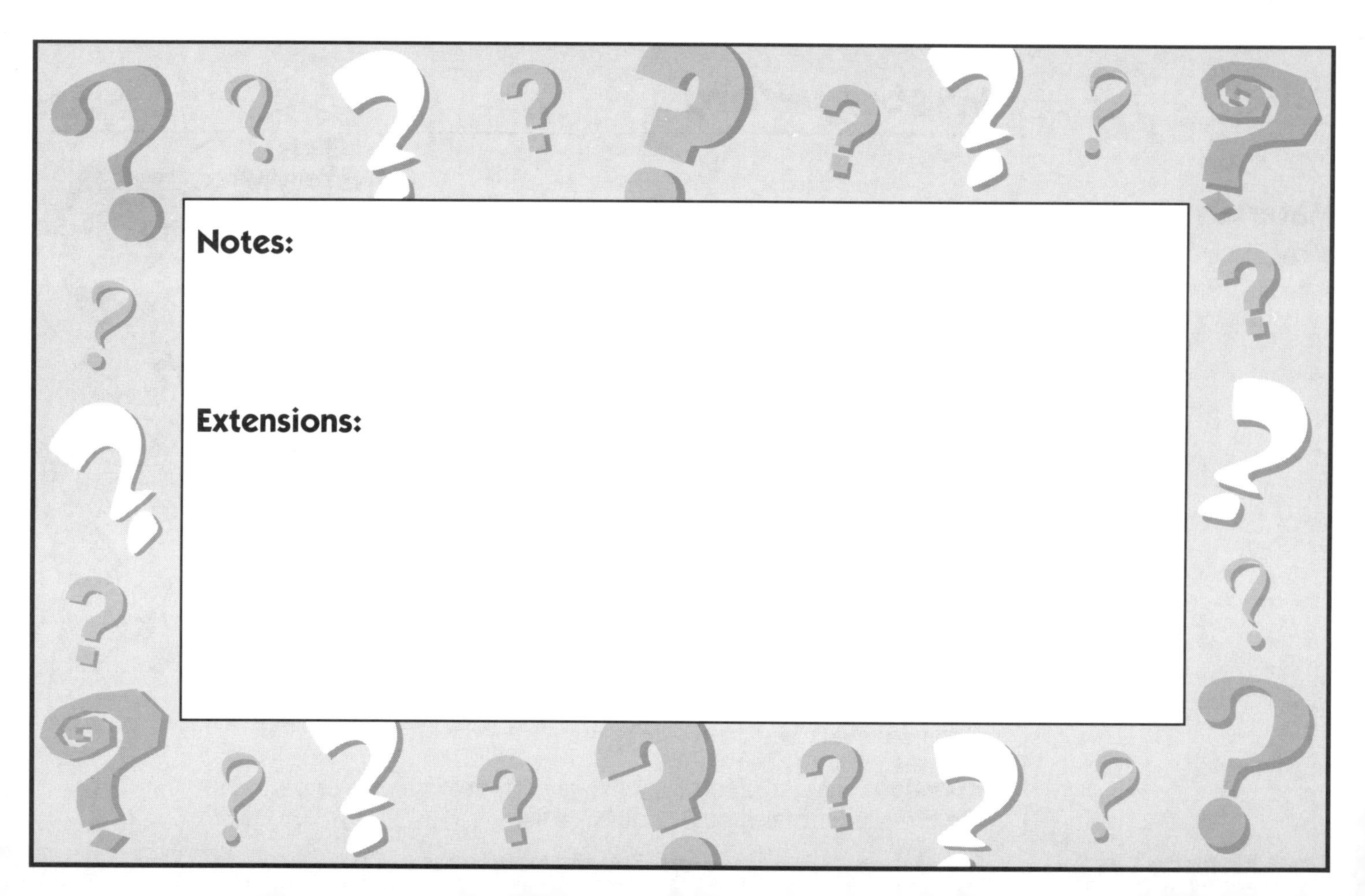

Notes:

Extensions:

What's My Rule?

Activity 19

Materials

- Attribute Blocks or a collection of objects for sorting

Procedure

1. Use this activity with a small group of students. Create a rule for sorting Attribute Blocks (e.g., all triangles, all red blocks, all small blocks). One by one, place blocks with that attribute in a separate set.
2. When a student thinks he or she knows the sorting rule, have him or her add a block without saying the rule aloud.
3. Continue until most students have selected an appropriate block, then have one student explain the rule.
4. Continue to sort in different ways, sorting by two or three attributes (e.g., large, thin blocks or red blocks with straight sides). Ask students to work in groups of two or three and repeat the activity.

Notes:
Extensions:

Class Sort

Activity 20

Materials

- collections of objects for sorting (beans, keys, buttons, pasta, nuts and bolts, pattern blocks)
- 12" x 18" construction paper
- markers
- index cards
- pencils

Procedure

1. Divide the class into partners, and give each student pair a collection of objects, construction paper, and a marker. Ask students to draw a Venn diagram on the construction paper.
2. Have students sort the objects by one or two attributes, placing them on the Venn diagram.
3. Tell students to write their sorting rules on index cards and place them face down by the Venn diagram.
4. Ask students to go around the room and try to determine the sorting rule for each Venn diagram. When they think they know the answer, have them turn over the index card and check.

Variation Label small paper squares with numbers *0–25*. Have students sort the numbers by one or two attributes.

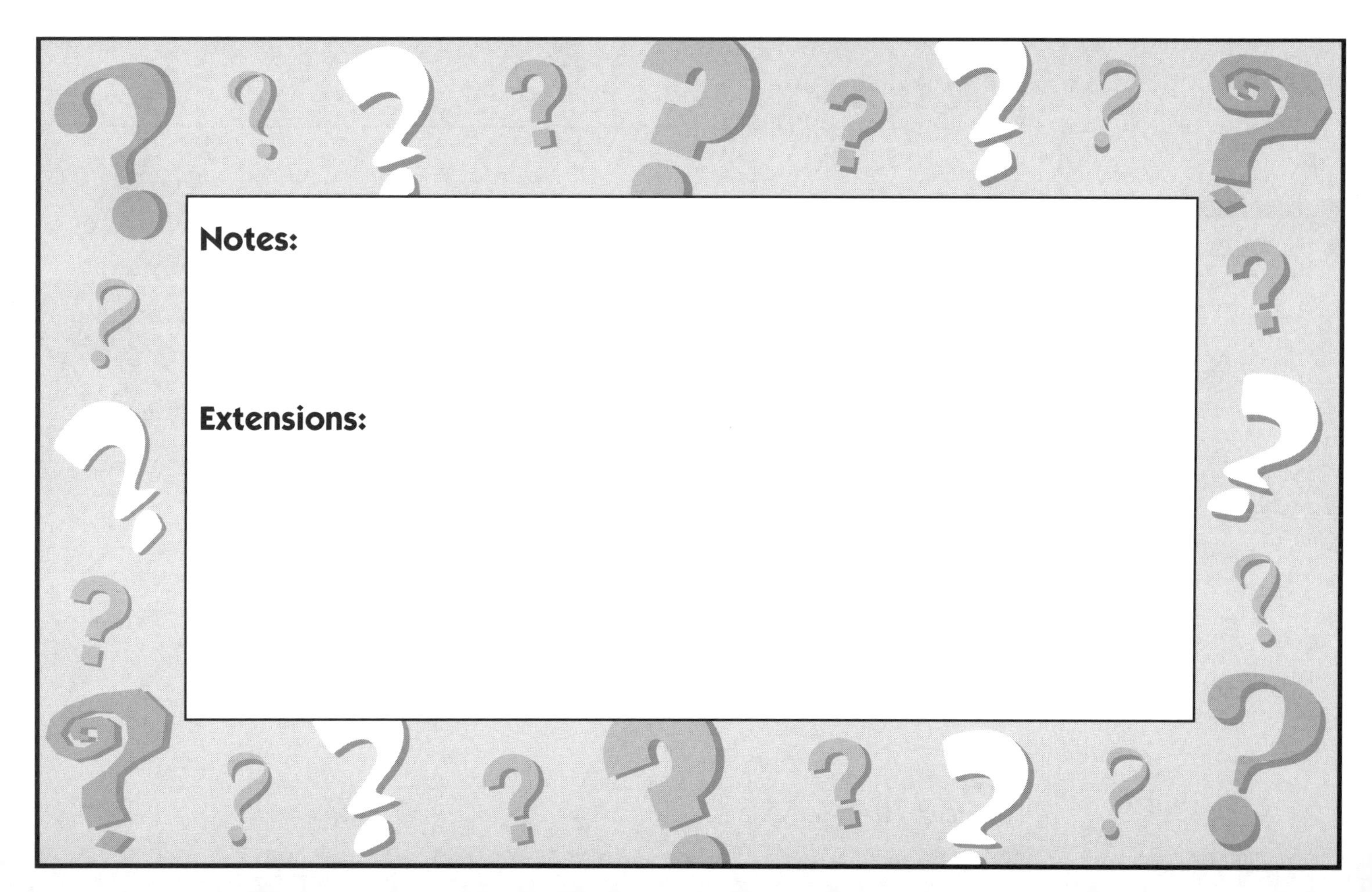

Notes:

Extensions:

Heads or Tails?

Activity 21

Materials

- coins
- pencils
- paper
- chart paper
- markers

Procedure

1. Ask students to predict how many times a coin would land heads up if they tossed it 50 times.
2. Have students work with a partner to toss a coin 50 times and record the results using tally marks.
3. When they finish, have students count how many times the coin landed heads up and how many times it landed tails up. Ask them to compare the results to their predictions.
4. Record partners' totals on a class chart, and have students compare results.
5. Show students how to write the results as a statement *(23 out of 50 landed heads up)* and as a ratio *(23:50)*.

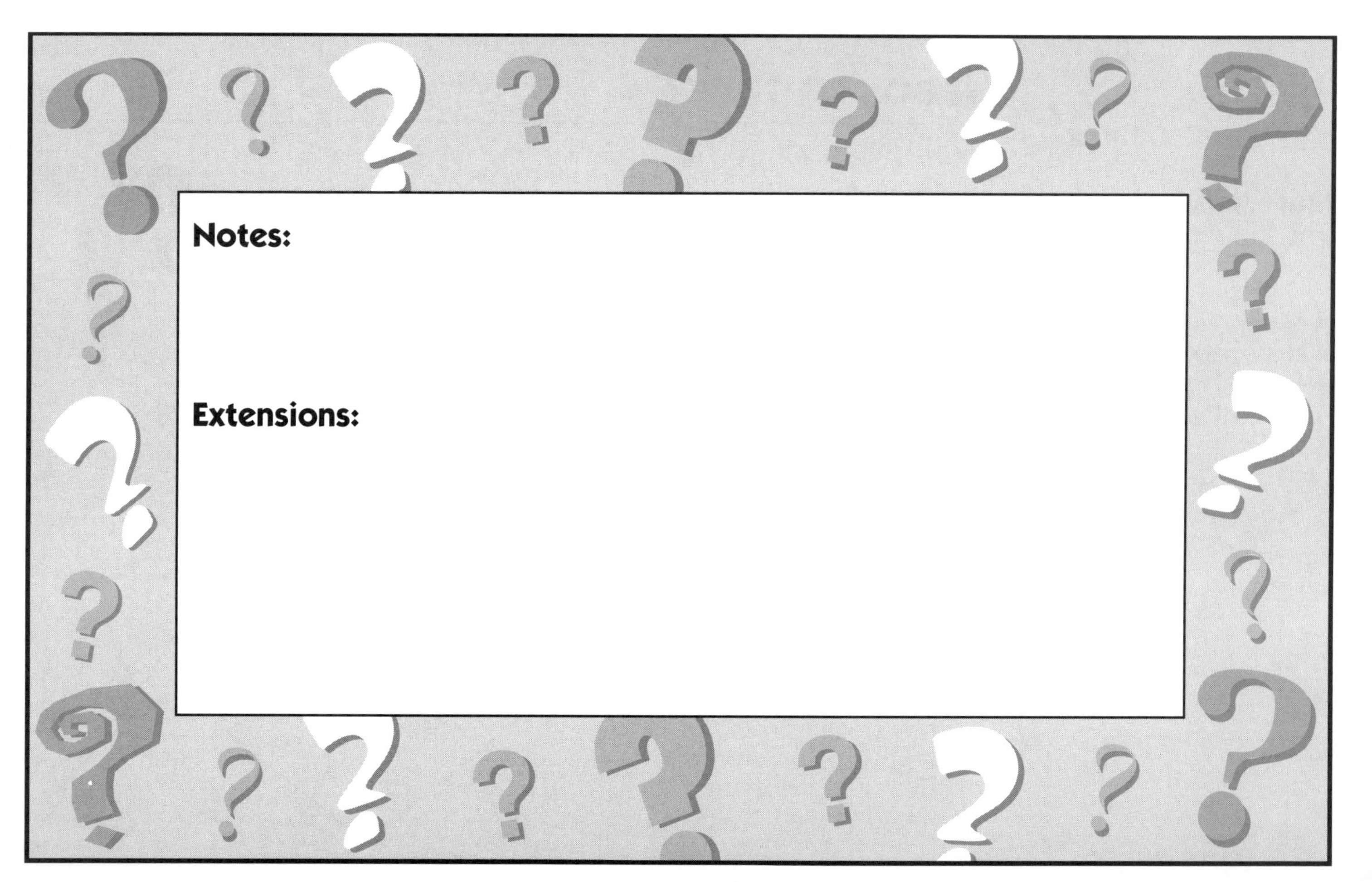

Notes:
Extensions:

Car Race

Activity 22

Materials

- 2–3 toy cars
- masking tape
- paper
- pencils

Procedure

1. After doing the *Heads or Tails?* activity (page 21), ask students if they think the odds are always the same when two possibilities are involved (i.e., "Do both have an equal chance?").
2. Using masking tape, mark a start and a finish line on a smooth floor surface.
3. Working in pairs, have students race the toy cars at least 20 times. Record the color of the winning car.
4. Have students think about how racing cars (people, boats, horses, and so on) is different than tossing a coin. Ask them to write several reasons for this difference. Is winning a matter of chance like tossing a coin? Why or why not? What else affects the results?

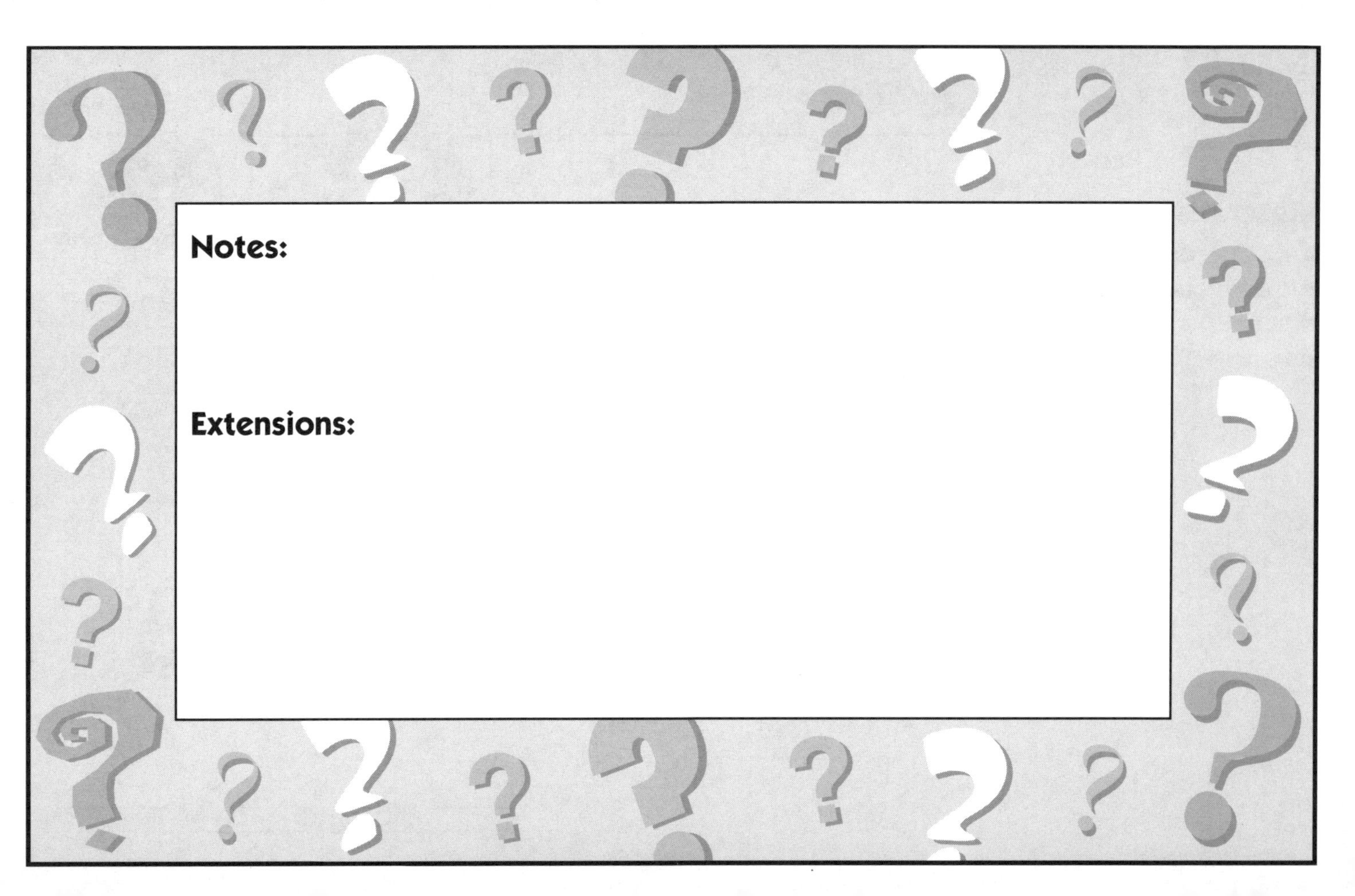
Notes:
Extensions:

Weather or Not

Activity 23

Materials

- newspapers
- paper
- pencils

Procedure

1. Have students listen to or check a local newspaper for the weather forecast every day for a week. Have them record the predicted and actual high and low temperatures for each day. Or, they can record a more general forecast (e.g., *clear, cloudy, chance of rain*).
2. At the end of the week, have students compare weather forecasts to the actual weather.
3. Ask students why they think weather forecasts are sometimes wrong or unpredictable.

Extension Ask students to research how weather forecasters arrive at their predictions.

Notes:

Extensions:

Letter Frequency

Activity 24

Materials

- paper
- pencils
- storybooks
- chalkboard
- chalk

Procedure

1. Have students work in pairs. Ask them to write the letters *A–Z* down the left side of a piece of paper.
2. Tell them to select two long paragraphs from a book. Have one student read aloud each letter of each word while the other student tallies the frequency of each letter.
3. Combine data from all students. Discuss which letters appear most/least often.
4. Play the traditional spelling game *Hangman.* Encourage students to use their new information about letter frequency to help them play the game.

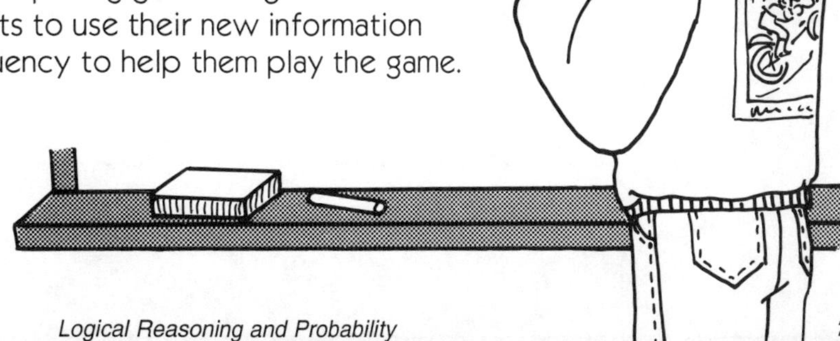

Notes:

Extensions:

Candy Consensus

Materials

- chart paper
- markers
- paper
- pencils

Procedure

1. Ask each student to respond to the following question by marking his or her preference on a chart: "What is your favorite candy bar—Snickers, Milky Way, or Hershey Bar?"
2. After tallying the survey results, have students make predictions about the results of a larger survey. Students could survey three or four more classes or select 100 children at random during recess.
3. Compile the data from the larger survey, and have students compare the results to their classmates' preferences.

Extension As a group, discuss how this activity compares to sample surveys taken by politicians and businesses.

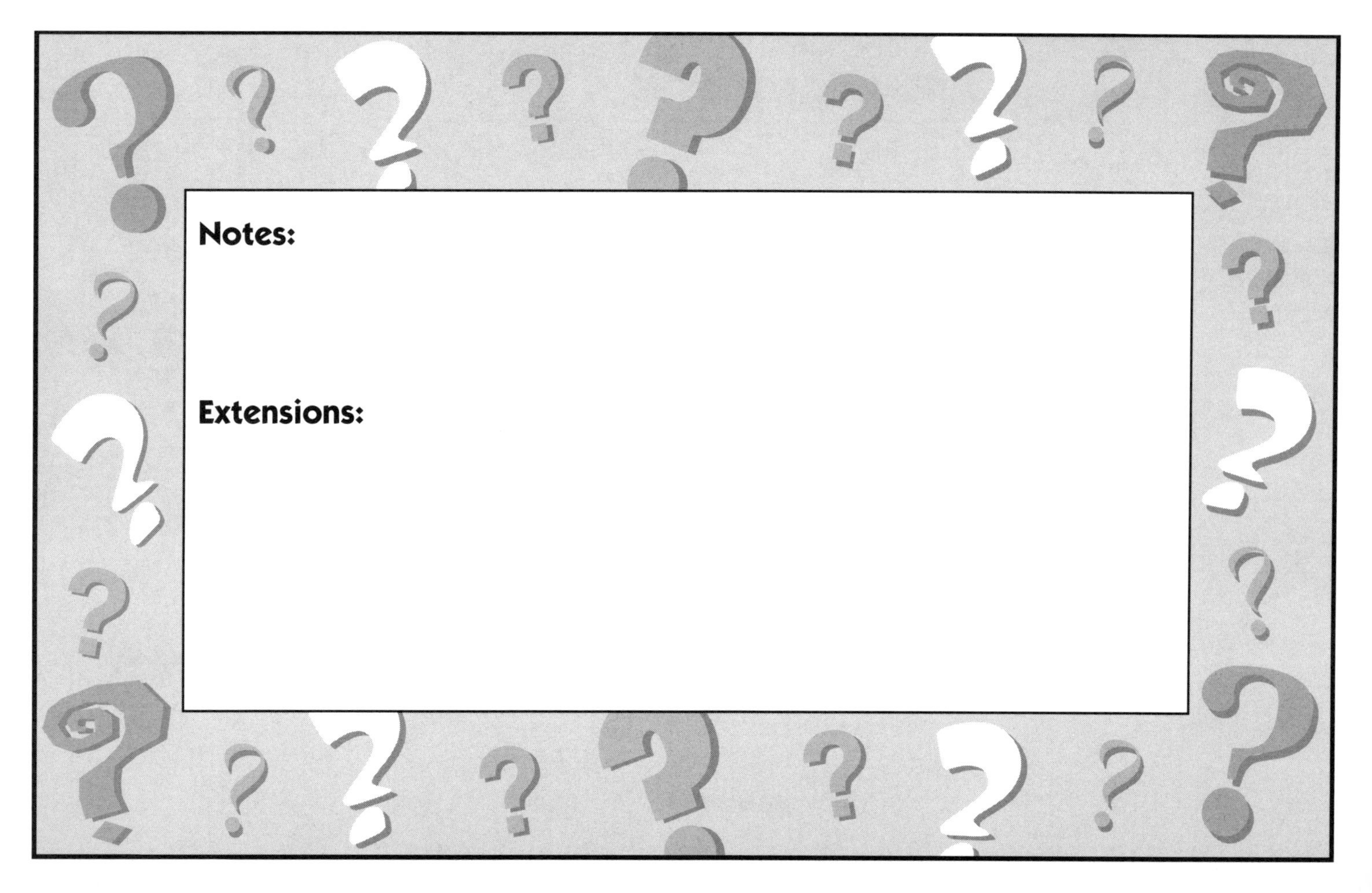

Notes:

Extensions:

Spinning Odds

Activity 26

Materials

- paper plates
- brass fasteners
- large paper clips
- crayons or markers
- paper
- pencils

Procedure

1. Give each student a paper plate, brass fastener, and paper clip. Have students divide the plate into four equal parts and color one-fourth blue, one-fourth yellow, and two-fourths red.
2. Have each student insert the brass fastener in the center of the plate and loop the paper clip over the fastener to make a spinner. Ask students the probability of spinning yellow or blue (one out of four), or red (two out of four).
3. Have students predict the outcome of 40 spins, then verify their predictions, tallying results as they work.
4. Ask students to study their recordings and discuss results. *What is the proportion of red spins to yellow or blue spins? Is there an equal chance of spinning yellow and red? Why or why not? How did your predictions compare to the actual results?*

Notes:

Extensions:

Do It Yourself

Activity 27

Materials

- paper plates
- brass fasteners
- large paper clips
- crayons or markers

Procedure

1. Tell students that two-to-one odds means the chance of one thing happening is twice as likely to occur as another. Ask students to design a spinner with two-to-one odds.
2. Give student pairs materials to make a spinner.
3. When spinners are complete, ask students to make up a game or a method for showing how their spinners produce two-to-one odds.

Variation Instead of step 3, have students explain with writing and/or drawings why their spinners would produce two-to-one odds.

Notes:
Extensions:

One in Six

Materials

- dice
- paper
- pencils

Procedure

1. Have students work in pairs. Give each student pair a die and some paper. Ask them to divide their papers into six sections and label the sections *1–6*.
2. Ask students if they think one number will appear more often than the others when rolling the die. If so, have each student write his or her name in the section with that number. (The chances of rolling any given number on a die is one in six.)
3. Ask students to take turns rolling the die and making a tally mark under the matching number.
4. After 36 rolls, have students add up the tally marks under each number.
5. As a group, discuss the following questions: *Was picking the winning number a matter of luck or skill? Did both students have an equal chance? Why or why not? If they played the game again, would the results probably be the same or different?*

Notes:

Extensions:

Probability Bingo

Activity 29

Materials

- paper
- pencils
- 2 dice
- game markers

Procedure

1. Have students fold a piece of paper into 16 equal sections.
2. Tell students they will be playing a bingo-type game. The numbers called will be determined by rolling two dice and adding the numbers together.
3. Have students write any number from 2 to 12 in each square. They can use each number as many times as they wish to improve their chances of winning.

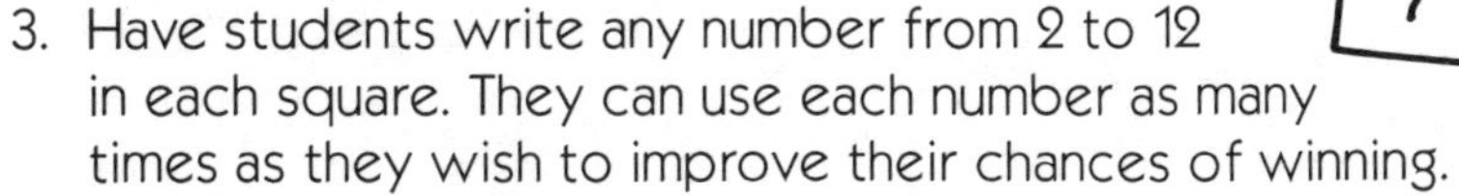

4. To play, have students take turns rolling the dice and calling out the sum of the numbers rolled. Tell them to cover one matching number with a marker. (Even if they have written a number several times, they can only cover one per turn.) The first person to cover four in a row wins.
5. Keep a tally of each sum called on the chalkboard so students can see the distribution of sums at the end of the game.
6. Play the game several times to see if the outcome changes. Allow students to erase and change numbers between games if they wish to improve their chances of winning.

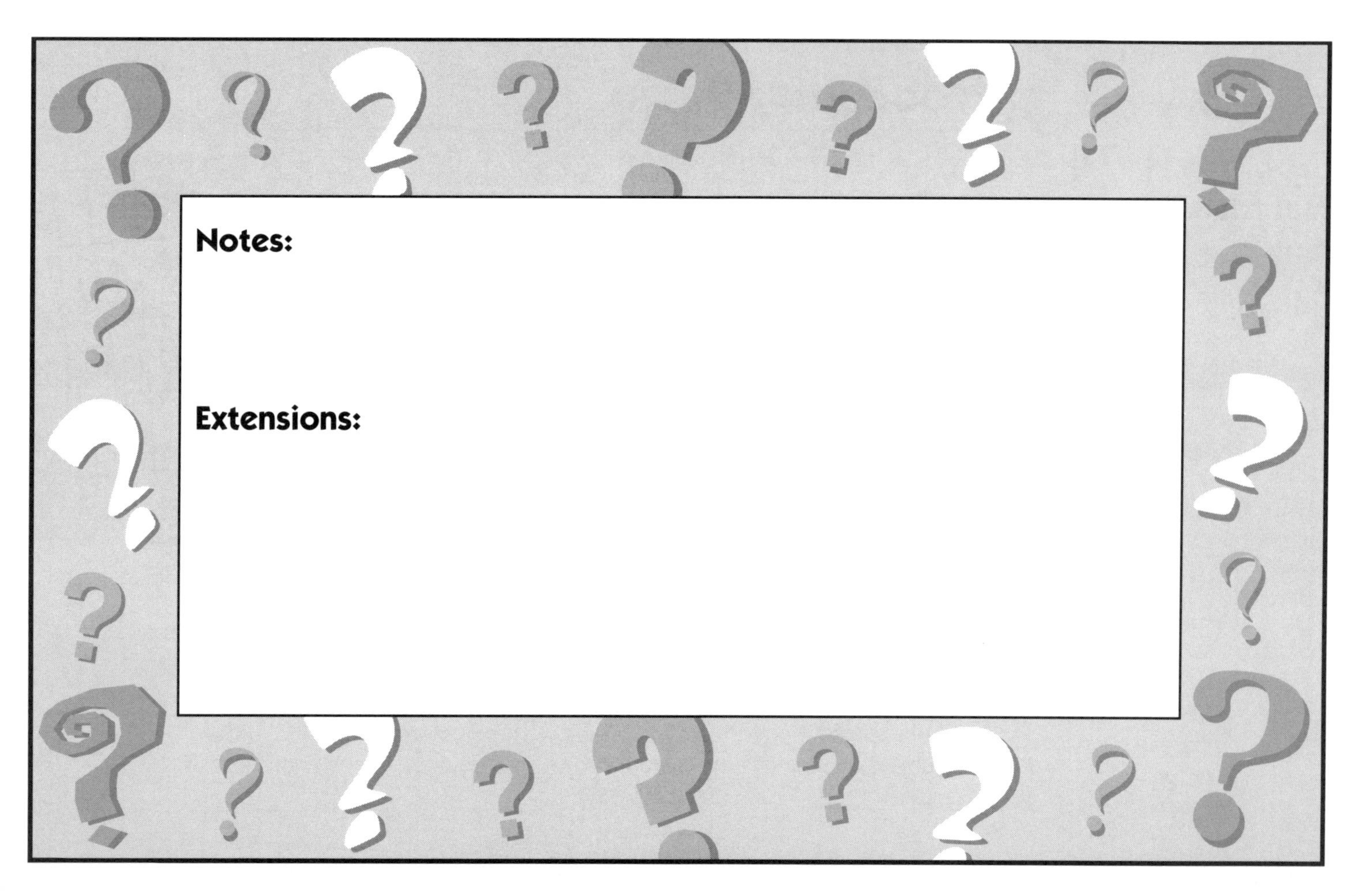

Notes:
Extensions:

Predict the Result

Materials

- 2 red, 2 blue, and 2 green counters, slips of paper, or beads
- paper lunch bag
- paper
- pencils

Procedure

1. Working with a small group, have one student place two red, two green, and two blue counters in the bag. Ask a student to choose one counter without looking in the bag. Have students predict which color will be chosen and why.
2. After everyone sees the color, set that counter aside. Based on the counters left, ask students to predict which color will be drawn next, and write down their choices.
3. Repeat step 2 with the remaining counters.
4. Have students compare their predictions to the actual results. Ask them how they revised their predictions after the first choice, after the second, and so on. How did the odds change each time a counter was removed?

Notes:

Extensions:

Three-Chip Choice

Activity 31

Materials

- 3 different-colored chips or counters
- small boxes
- paper
- pencils
- graph paper
- crayons or markers

Procedure

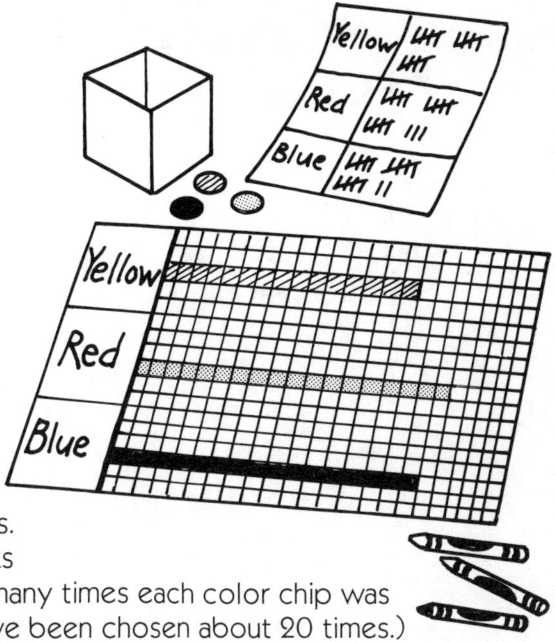

1. Give each student a piece of paper and have him or her divide it into three rows. At the left of the rows, have students write the three chip colors.
2. Divide the class into partners. Give each student pair a box and three different-colored chips. Ask students to take turns selecting one chip from the box with eyes closed, recording its color with a tally mark, and returning it to the box. Ask students to do this 60 times.
3. Have students add up their tally marks and make a bar graph showing how many times each color chip was selected. (Each color chip should have been chosen about 20 times.)
4. Have students compare their results. Discuss the odds of picking any one color chip (one in three).

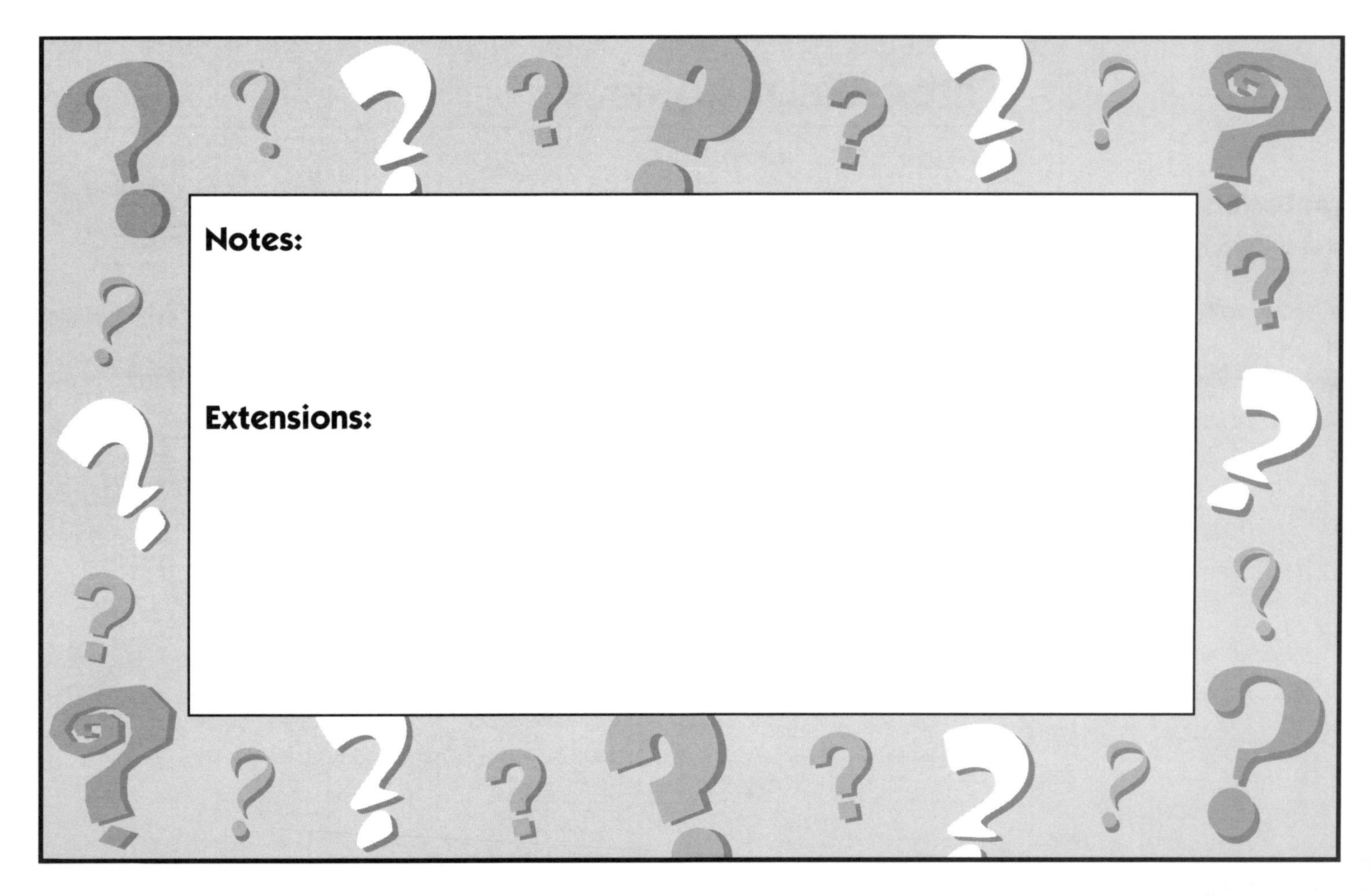
Notes:
Extensions:

Chip Ahoy!

Activity 32

Materials

- 100 chips or counters (75 of one color and 25 of another)
- paper lunch bag
- paper
- pencils

Procedure

1. Place the chips in the bag. Working with a small group, tell students they will take samples to help determine approximately how many chips of each color are in the bag.
2. Have students divide their papers in half and write the name of each color at the top of each column.
3. Ask one student to draw 10 chips from the bag with eyes closed. Ask students to record the number of chips of each color, and then return the chips to the bag.
4. Repeat step 3 until all students have had a turn. Add the totals for each color.
5. Tell students there are 100 total chips in the bag. Using their results, ask students to estimate how many chips of each color are in the bag. Have them write or explain how they came up with their estimates.
6. Reveal the actual numbers, and have students compare them to their predictions.

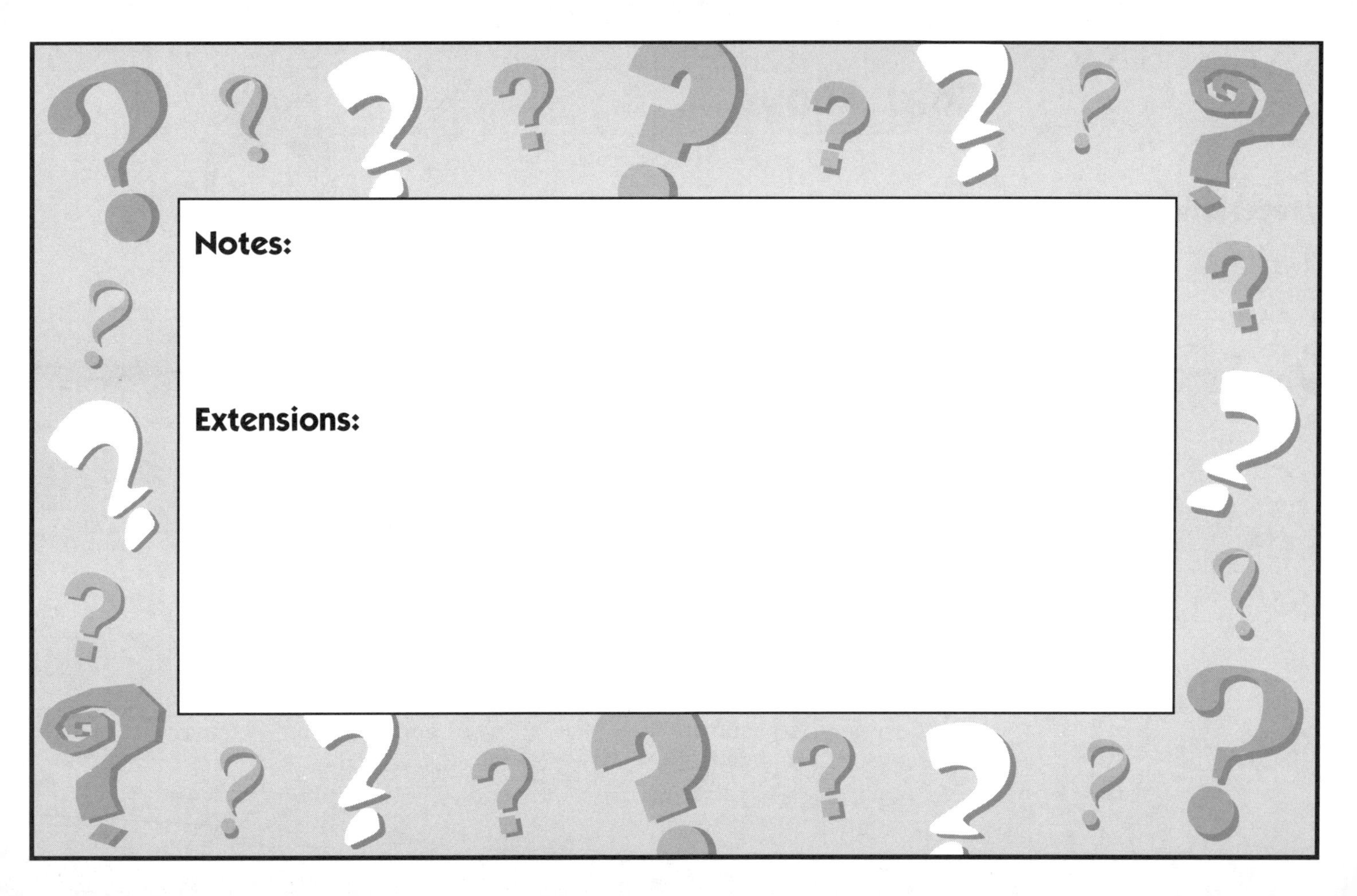

Notes:

Extensions:

How Many Possibilities?

Activity 33

Materials

- pennies
- paper
- pencils
- chart paper
- markers

Procedure

1. Divide the class into partners. Give each student pair six pennies and ask them to find all possible combinations of heads and tails that could result from tossing all six pennies.
2. Students can record one possibility on each line by using *H* for heads and *T* for tails (e.g., *H H H T T T*). Encourage students to manipulate the pennies to find various combinations.
3. With six pennies there are seven possible combinations. Ask students who did not find all seven to go back and try again.
4. Ask partners to toss the pennies 25 times and record the data.
5. Combine class data on a chart. Ask students for the least likely combinations.

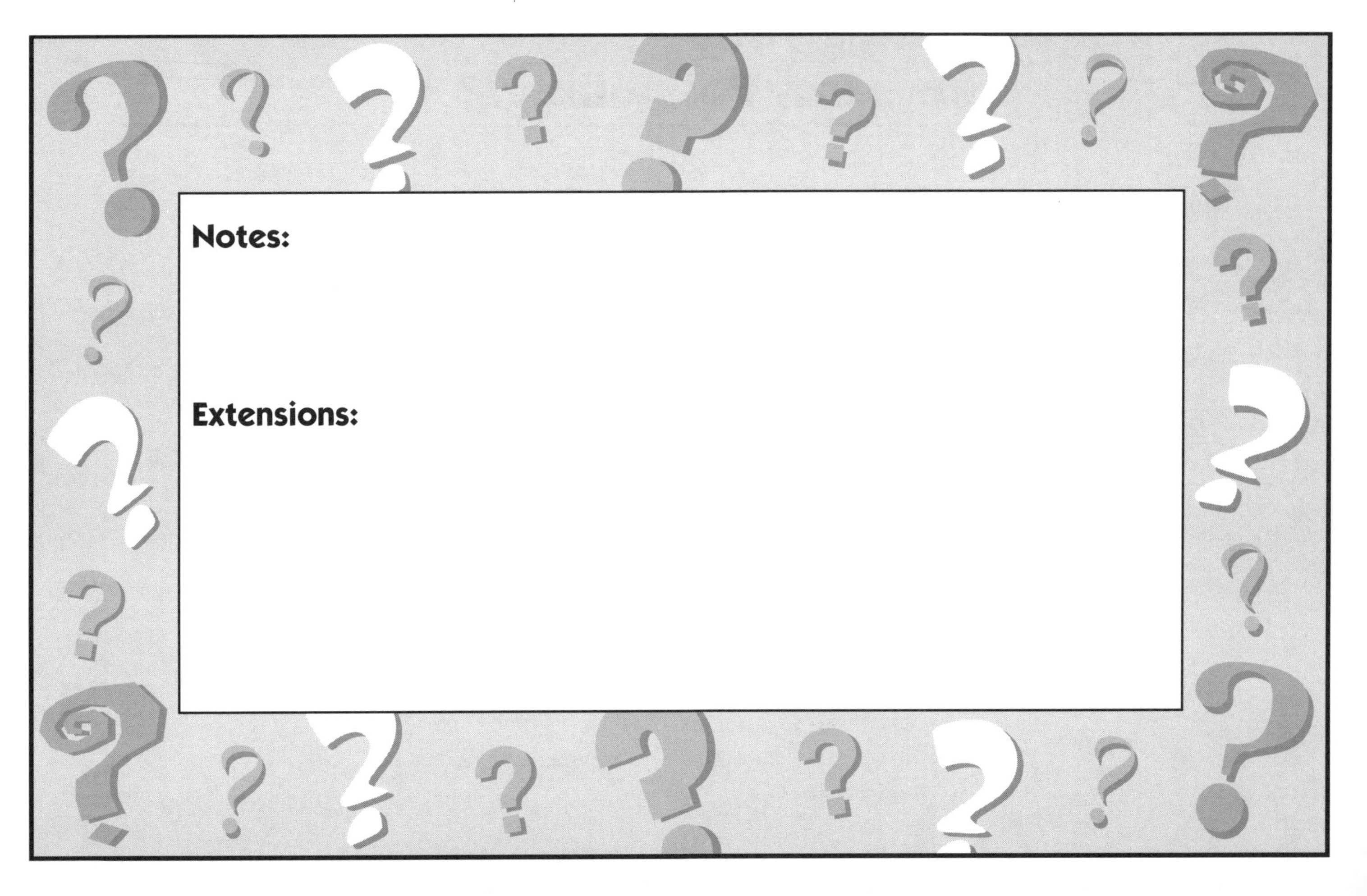

Notes:

Extensions:

Is It Fair?

Activity 34

Materials

- dice
- blank cubes
- fine-tip marker
- paper
- pencils

Procedure

1. Label two sides of the blank cubes *1*, two sides *2*, and two sides *3*.
2. Give each student pair a die and a cube.
3. For the first game, have partners roll the cube. Player A gets a point if the number is even. Player B gets a point if the number is odd.
4. Have students roll a total of 30 times, tallying the outcome of each roll. Ask students if they think the game is fair or unfair and why.
5. For the second game, tell students to take turns rolling the die. Player A gets a point if the number is even. Player B gets a point if the number is odd. Ask students if they think this game is fair or unfair and why. (There are three even numbers and three odd numbers on the die.)

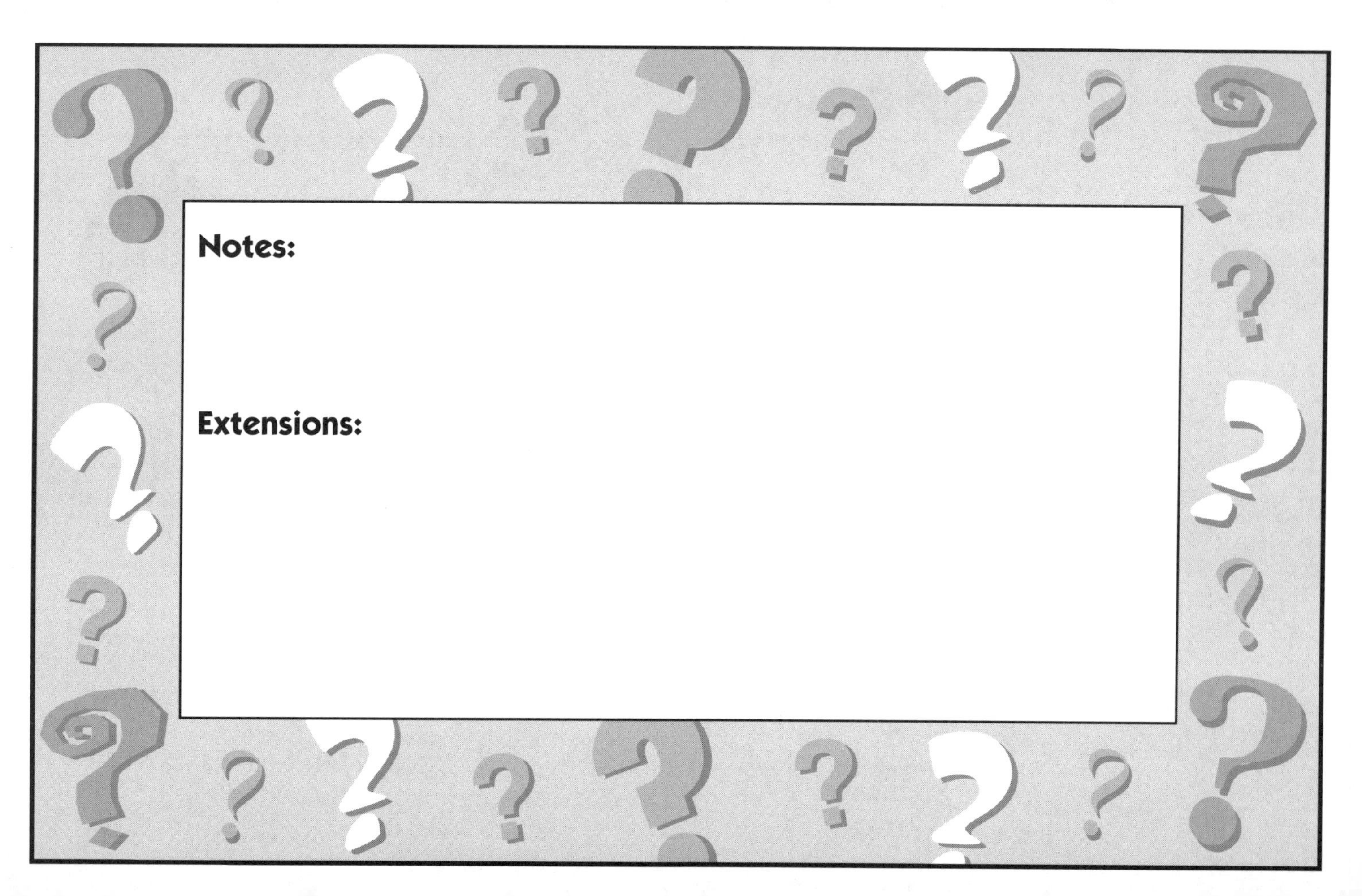

Notes:

Extensions:

Match It!

Materials

- dice
- paper
- pencils

Procedure

1. Divide students into partners, and give each student pair two dice.
2. Have students predict how many times out of 25 throws they will match their partner's number on a die. Ask students to write their predictions without looking at their partner's predictions.
3. Have partners simultaneously roll their dice and record their results as a *Match* if they get the same number or as *No Match* if they get different numbers.
4. Have students compare their predictions to the results after 25 throws.
5. Ask students to revise their predictions based on the results of the first game and play again. Did their predictions come closer the second time?

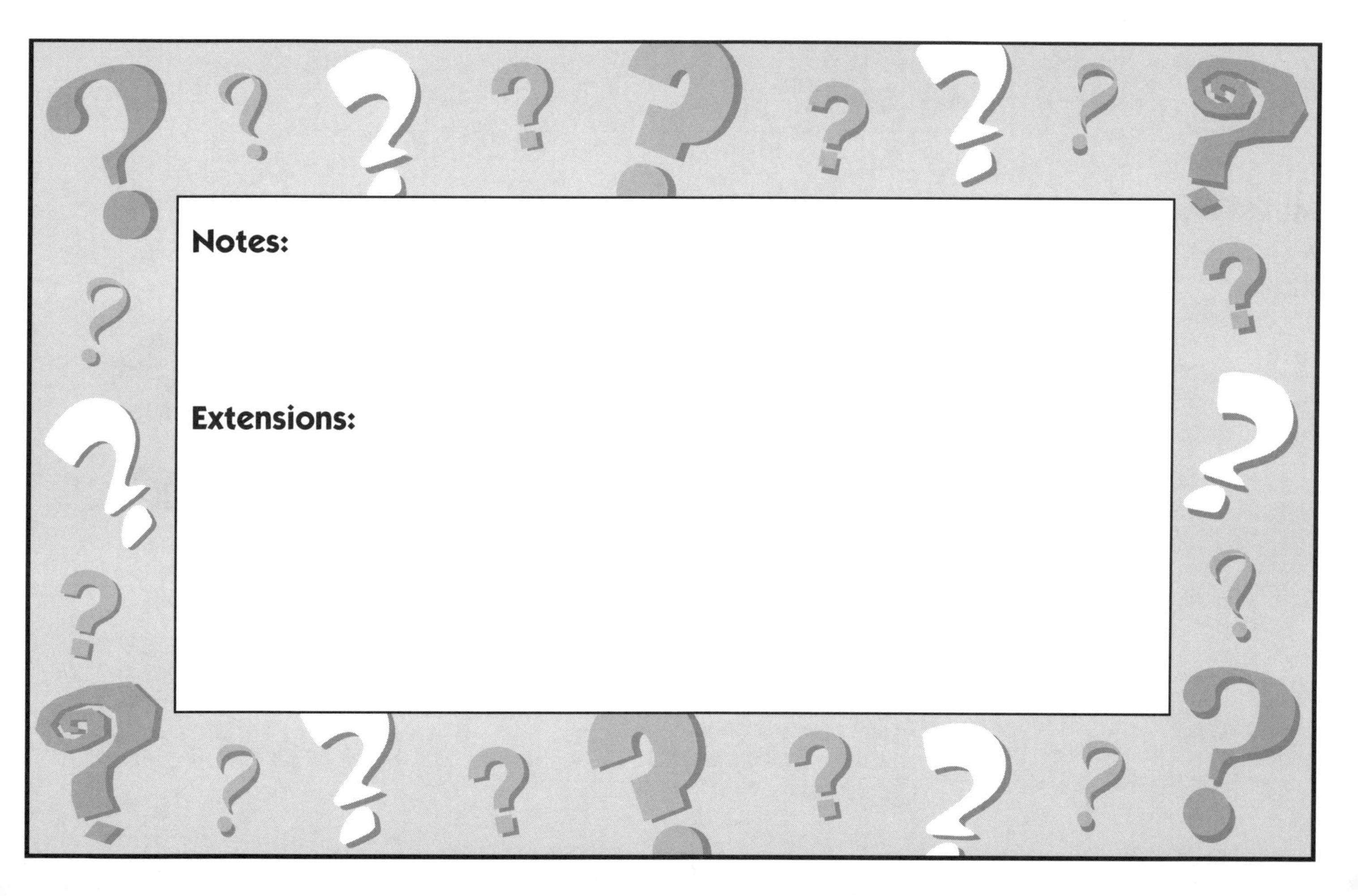

Notes:

Extensions:

Parent Letter

Dear Parents,

In our child-centered mathematics program, we are exploring the concepts of logical reasoning and probability. We will use a variety of materials to make the activities both interesting and educational. Some of the items are consumable. If you would be willing to purchase an inexpensive item for classroom use, please sign and return the bottom portion of this letter. I will then return the form to you stating the item and the date it is needed. Thank you!

Sincerely, ______________________________

- -

Yes, I would like to contribute an item for logical reasoning and probability activities.

____
(signature)

Please send in ______________________________ by ____________________.

Thank you very much!